MW01701429

Crocodiles and Alligators

Special Edition
Created by Angel Entertainment Inc.

Published in the U.S. by
Marboro Books, a division of
Barnes and Noble Bookstores, Inc.

Library of Congress Cataloging-in-Publication Data

Farré, Marie

(Young Discovery Library; 1)

Translation of: Qui a peur des crocodiles?

Includes index.

Summary: Describes the physical characteristics, behavior, and peculiar habits of crocodiles, including how to distinguish them from alligators.

1. Crocodiles — Juvenile literature. 2. Alligators — Juvenile literature. [1. Crocodiles] I. Wallis, Diz, ill. II. Title III. Series: Young Discovery Library (Series); 1.

QL666.C925F3713 1988 597.98 87-31804

ISBN 0-944589-01-4

Printed and bound by L.E.G.O., Vicenza, Italy

Written by Marie Farré
Illustrated by Diz Wallis

Specialist Adviser:
Dr. Donald Bruning
New York Zoological Society

ISBN 0-944589-01-4
First U.S. Publication 1988 by
Young Discovery Library
217 Main St. • Ossining, NY 10562

Translated by Sarah Matthews

Thanks to Aileen Buhl

YOUNG DISCOVERY LIBRARY

Crocodiles and Alligators

Once upon a time,
crocodiles fed on cakes and honey...

YOUNG DISCOVERY LIBRARY

In ancient Egypt, crocodiles were worshipped as gods. At the time of the Pharaohs a whole city, called Crocodilopis, was built in their honor. The crocodiles lived in a sacred lake. Their priests put gold bracelets on their arms, and fed them on cakes and honey.

The crocodile god was called Sebek.

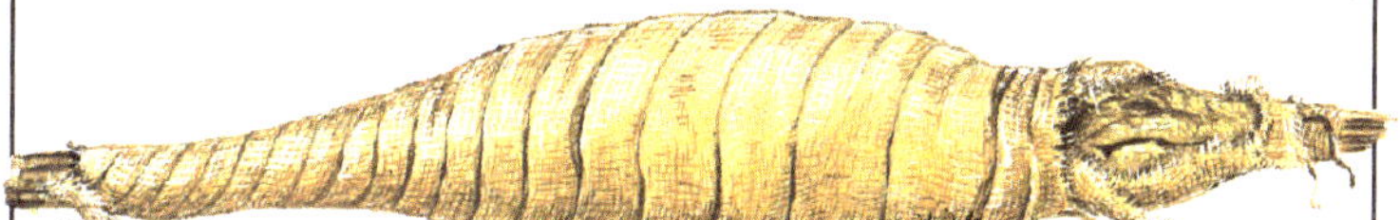
Crocodile mummy

When a crocodile died, the priests mummified it. They dried the body and wrapped it in herbs and bandages so that it did not rot.

Have you seen a crocodile?
They don't look very fierce lying in the zoo. But in the wild no creature dares attack a full-grown crocodile — except man. They live beside lakes and rivers in Africa, Asia, Australia and parts of America.

Crocodiles are reptiles…

like snakes…
tortoises…
lizards.
They have dry, scaly skin and lay eggs on land.

They are the most powerful reptiles in the world.
Crocodiles have lived in the world since the time of the dinosaurs.

A time when reptiles ruled the world.

Crocodiles have several close relatives:
alligators,
cayman, gharials.

Nile crocodile

How can you tell if it's a crocodile?
By the fourth tooth which sticks out on each side of the lower jaw.

Or a cayman?
They have a black crest between their eyes.

Central American cayman

Or an alligator? Their heads are broad, and their fourth tooth does not stick out.

Or a gharial?
By its long, narrow jaw, which helps it
catch
fish
to eat.

◆ Mississippi alligator

Indian ◆ gharial

The noise a crocodile makes is called a bellow, though it sounds more like a groan. Did you know that there are some crocodiles which are four times as big as you are, and over forty times heavier?

A crocodile's skin is as tough as armor.

It is made of scales. But it's not stiff like tortoiseshell. The scales fit beside each other and the skin moves and bends easily.

A crocodile has sharp claws on its feet. The scales on its tail stand up to form a spiky ridge.

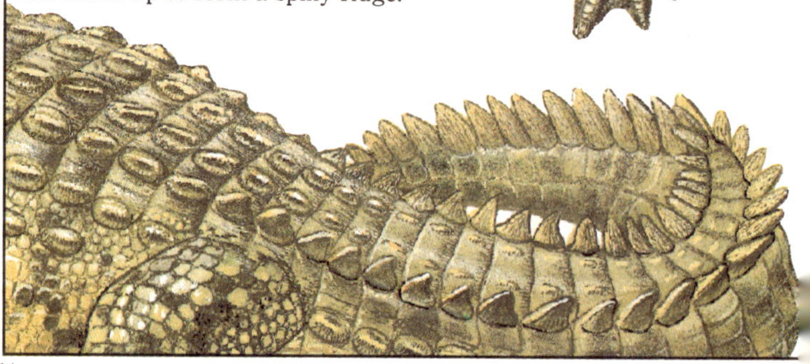

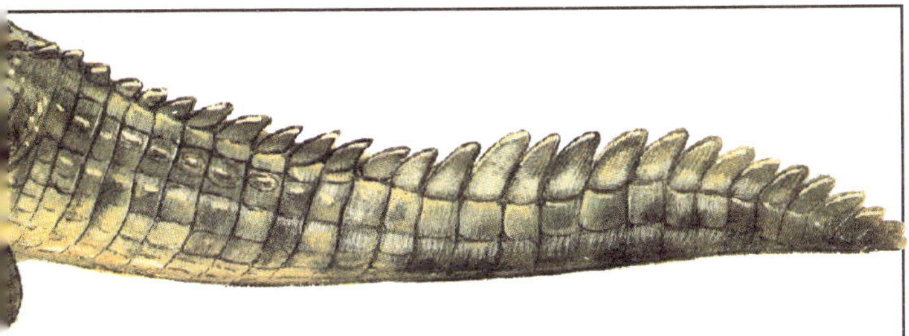

A crocodile can see at night like a cat. The pupils of its eyes get wider in the dark.

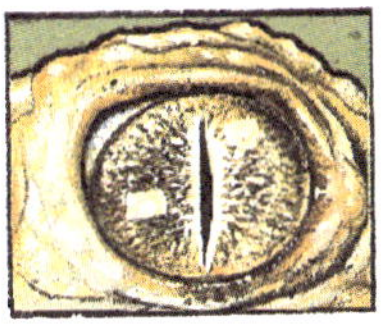

In daylight

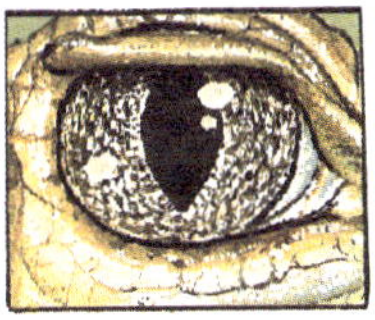

At night

What are its weapons?

Its powerful tail and its strong jaws. A crocodile has a lot of teeth. They come out easily, but when one falls out, another one grows in its place straight away. A tooth can be renewed 45 times.

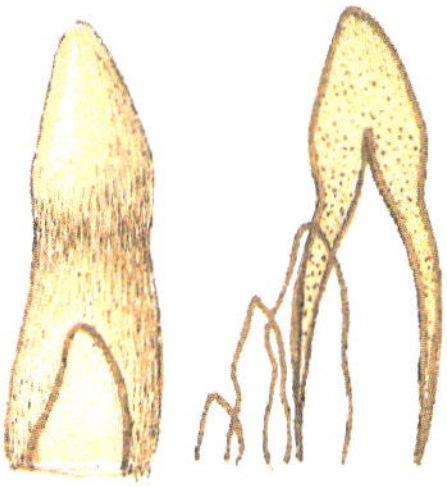

They have a piece of stretchy skin at the corners of their mouths so that they can open them very wide.

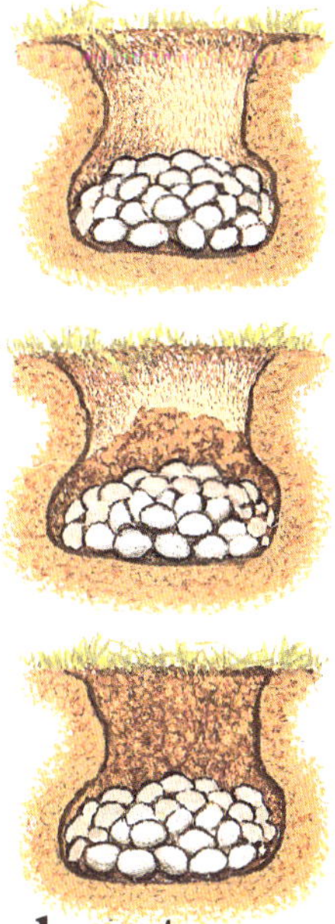

The nest is shaped
like a deep basket.

How are crocodiles born?

In the spring, the mother crocodile digs a nest with her front legs. In the nest she lays about fifty leathery white eggs, the size of hens' eggs. Then she covers the nest with sand.

Watch out! A lot of animals love to eat crocodiles' eggs!

While the eggs are incubating, the mother hardly leaves the nest. If the weather is hot, she scurries down to the water for a dip, then lies over the nest to cool it down.

Monitor lizards are very crafty. While the male lures the crocodile away from the nest. the female eats the eggs.

When the eggs are ready to hatch, after nine or ten weeks, the baby crocodiles call out from inside the eggs. While their mother scrapes the sand from the nest, they break out of the eggs, using a special **egg tooth** which drops off after they have hatched.

A baby crocodile is the size of a small lizard.

The dash for the water

As soon as they are out of the egg, the babies head for the water. They know how to swim immediately! Sometimes their mother carries them down to the water in her mouth.

Baby crocodiles eat small worms, snails and insects.

Do all baby crocodiles grow up?

One baby cayman the less...

Jaguars, storks and even bigger crocodiles enjoy eating them. So the young crocodiles have a special nursery area, with holes in the riverbank to hide in, where they can grow up safely. Despite all this care, though, only about two crocodiles from each clutch of eggs survive long enough to lay eggs themselves. But these two may live at least fifty years.

At 7 years old, it is an adult.

A crocodile swims by moving its tail. Only its eyes, ears and nostrils show above the water!

Why do crocodiles sleep in the water?

Like all reptiles, crocodiles lose their body heat when the temperature goes down. But water keeps the sun's heat longer than the air, so crocodiles stay all night in the warm water and only crawl out at dawn when the water is becoming cooler, and the sun has risen. They lie in its warmth on the bank.

A crocodile can stay underwater for up to two hours.

It can close its nostrils, then the water cannot get in. It has a special transparent eyelid so that it can see underwater. Its ears are protected underwater too, by a thin skin which stops water getting in, but still lets sound through.

What animals do crocodiles attack?

Big animals such as buffalo, zebra, gazelle, smaller animals like hyena — sometimes they even attack elephants, but they only manage to bite their noses. When they get old, crocodiles may be eaten themselves, by other crocodiles.

Do they eat often?

A crocodile can go two or three months without eating if it has to.

A dozen crocodiles can kill and eat a hippo.

Crocodilcs are cunning hunters.
You can see how a crocodile hunts a gazelle. After lying in ambush, it knocks the gazelle into the water with its tail, and jumps in after it, catching one of its legs in its mouth. It pulls the gazelle under and drowns it. It tears bits off by twisting over and over. The pieces must be swallowed whole as a crocodile can't chew.

Did you know that a crocodile swallows pebbles?
It may do this because the small stones help to break up the food which the crocodile has gulped down.

After it has eaten, the crocodile will lie and rest on the bank in the sun. If it gets too hot it will open its mouth and let the evaporation from the damp skin inside cool it down. Or it might burrow into the cool mud.

How do crocodiles move on land?
They crawl or trot, or sometimes, on wet grass, they glide, pushing themselves along with their tails.

If, one day, you find yourself being chased by a crocodile, run in zigzags. Crocodiles can't turn easily, and this slows them down.

Two crocodiles snort warnings at each other.

Each group has its own territory.

Crocodiles usually live in groups. The boundaries of their territory are marked by the males with a stong smell of musk, produced by a gland under their bellies.

They bite each other and thrash with their tails.

Everyone knows his place.

The leader has the best spot on the riverbank, the greatest number of females, the biggest bits of meat. Old crocodiles lie on the soft grass, the younger ones have to make do with the sloping banks.

The loser lifts his muzzle to show he's beaten, and then makes off.

Do crocodiles really cry?

If you pretend to be sad, people may say you're crying crocodile tears. But crocodiles only have tears in their eyes if they yawn, or, if they've been in the sea, to wash the salt out.

Crocodiles have the strangest toothbrushes!

They are birds: **plovers.** They peck off the leeches and the bits of food that get stuck between the crocodile's teeth. **Spur-winged plovers** help the crocodile in a different way by eating the ticks that live between his scales.

When danger comes near, the plovers cry out, and the crocodiles all dive underwater.

These animals, on Earth since prehistoric times, are dying out. Man has almost destroyed them. Their skins are valuable, and they used to be hunted for them, by men with guns. Nowadays we are beginning to realize that crocodiles are precious and irreplaceable. They are protected and live in reserves. The skin used for expensive handbags, belts and shoes comes from crocodiles raised on special farms. The musk is used in perfumes.

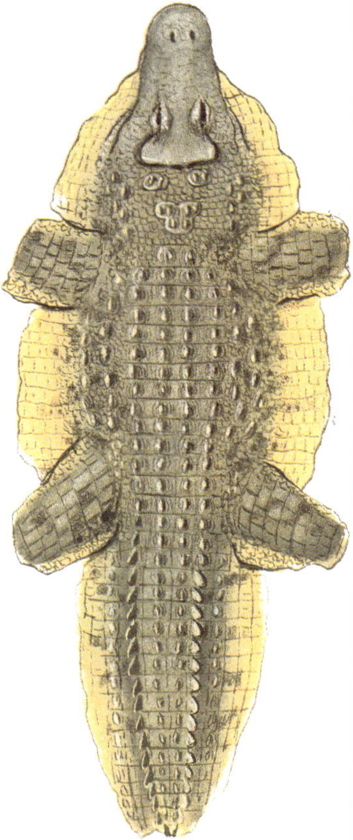

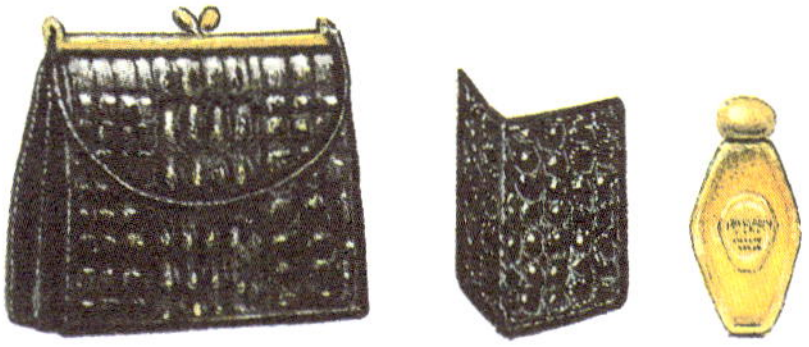

These products may be beautiful — but crocodile skin looks better on crocodiles.

Index

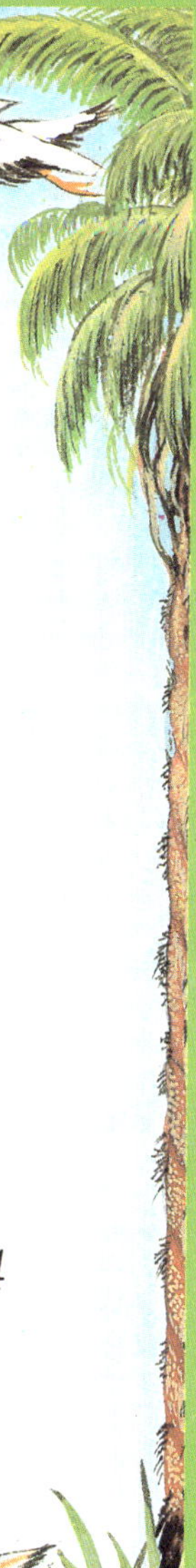

The Crocodile

How doth the little crocodile
   Improve his shining tail,
And pour the waters of the Nile
   On every golden scale!

How cheerfully he seems to grin!
   How neatly spread his claws,
And welcomes little fishes in
   With gently smiling jaws!

Lewis Carroll

Books of Discovery for children five through ten...

Young Discovery Library is an international undertaking — the series is now published in nine countries. It is the world's first pocket encyclopedia for children, 120 titles will be published.

Each title in the series is an education on the subject covered: a collaboration among the author, the illustrator, an advisory group of elementary school teachers and an academic specialist on the subject.

The goal is to respond to the endless curiousity of children, to fascinate and educate.

TITLES IN THIS SERIES:

The Barbarians

Odile Bombarde

Here's a complete introduction to the tribes that came to conquer and stayed to settle Europe in the Middle Ages—featuring pictures and descriptions of Attila, the Vikings, Charlemagne and more.

Crocodiles and Alligators

Marie Farré

The daily lives, evolution, eggs, babies and other natural history of the big reptiles from the time they were worshiped in ancient Egypt to today—plus a poem by Lewis Carroll.

Monkeys, Apes and Other Primates

André Lucas

Entertainment and education both abound—as these realistic renderings show dozens of species and engaging text tells why lemurs stay up all night, gorillas are strong, gibbons fly—and more!

Going West: Cowboys and Pioneers

Martine Courtault

Adventures of the pioneers, including details of a two-month trip across the Great Plains, life during the Gold Rush, settling a new town, the Pony Express and everyday life in the old west.

Bears, Big and Little

Pierre Pferrer

Are bears as clumsy as they seem, what do grizzlies eat, how do mother bears raise their young? This book answers these and other important questions as it describes all eight species of bears and their habitats.

Long Ago in a Castle

Marie Farré

Realistic rendering of many castles across Europe open this book-full of facts about castle communities of the Middle Ages—replete with information on building techniques, home life, knights, feasts and lords.

Following Indian Trails

Nicholas Grenier

Information to defy stereotypes and teach how Native Americans came to this land, acquired horses, built villages, taught children about life and protected their land.

Undersea Giants

Patrick Geistdoerfer

Poetry, lore, authoritative renderings and text join in this introduction to whales, seals, dolphins and other sea mammals—with many fascinating facts about communication, reproduction and protection of the species.